Quantum Physics

and

Other Life Lessons

Quantum Physics

and

Other Life Lessons

Jonathan Warren

RESOURCE *Publications* · Eugene, Oregon

QUANTUM PHYSICS AND OTHER LIFE LESSONS

Resource Publications
An Imprint of Wipf and Stock Publishers
199 W. 8th Ave., Suite 3
Eugene, OR 97401

www.wipfandstock.com

PAPERBACK ISBN: 978-1-6667-1649-8
HARDCOVER ISBN: 978-1-6667-1650-4
EBOOK ISBN: 978-1-6667-1651-1

SEPTEMBER 22, 2021

Dedicated to Mother Goose, Mrs. Ann Story.

Contents

Life Lessons

Quantum Physics

GRAVITY

How boring
it would be,
if everything
in our universe
could be
explained,
summed up
with diagrams
drawn
on
dusty chalkboards
by pale fingers,
and how
exciting
that
there are
still things
scientists
can't explain,
and it is
this unknown,
unproven,
untested piece
of our world
that gets me
out of bed
in the morning,
to investigate
the ends
of rainbows,
turn over
rocks,
search
for monsters,

and battle
windmills.
It keeps
my dream
of flying
alive
because
gravity
is just
a rule
and
rules
were made
to be broken.

RELATIVITY

Relativity

JW

I walk in circles
but that's alright,
because the universe
is curved
and my steps
roll
like a marble
around a snail shell,
and my thoughts
escape upwards
into the heavens
to join the rest.
I walk in circles
but that's alright,
because I know
time bends
and curves
around me,
to before
the dam broke,

before
the flood
washed
her away,
and
I know
that we
aren't apart,
because the space
between us
is granular,
so there
is no space
between,
and I am holding her
right now,
so
I don't think
of my loneliness
instead,
I imagine
a black hole
imploding
on itself,
sucking gravity
into
another dimension,
as I stand here,
with a bucket
and a mop,
waiting
for the sky
to fall.

LIGHTYEARS

I sit here
beneath the sun,
soaking in
it's energy
as it travels
the lightyears
between
here and there,
unknown
to known,
cold
to warm.
I'll absorb
all the knowledge
I can
about
how to feed
the living
and
honor
the dead,
carried
in a ray
of light,
which isn't
a solid ray
at all
but thousands,
millions,
billions
of light particles,
molecules
travelling together,
side by side,

one path,
one purpose.
To the outsider,
it appears
as a beam
and that is what
our minds
like to think,
need to think,
but
nothing is real,
nothing is solid.
Matter,
is made
of tiny bits
of our universe,
conversing,
travelling together,
for a minute,
an hour,
a blink,
a lifetime,
forming temporarily
into the chair
you think
you are sitting in,
that will
turn to splinters
upon the grass
you can't prove
is under
your feet,
beneath
a ray
of sunlight
that isn't
warming
your skin.

Science Fair

Do black ants
prefer
brown sugar
or white,
begged
the thesis
of my third grade
science fair project.
I conducted
the experiment
flawlessly,
hypothesis intact
my results
were conclusive.
Black ants
don't care
if the sugar
is brown
or
white,
they prefer
to follow
the ant
directly
in front
of them.

THERMODYNAMICS

Thermodynamics

and Carlo Rovelli
tell us
that without heat,
things remain
unchanged
and the future
behaves exactly
as the past,
but there
is no such thing
as heat,
and it can't
be measured
in caloric fluid
because,
heat is an illusion
caused by particles,
which are moving
quicker
than "colder" ones.
Scientists wonder
if a hot body
could become

warmer
through contact
with a colder one,
Einstein wondered
about a box
filled with light,
and I wondered
about you,
before
the particles
between us
slowed,
as we burned,
our energy
could fill
a box of sky.
I searched the city
for you,
but when
found
you had
turned cold,
when I touched
your hand,
I felt your chill
creep
into my limbs,
and I knew
you would remain
unchanged.

Approximation

They say
it's not
the years,
it's the miles,
but the theory
of Relativity
argues
that it's not
the miles,
it's the height.
One twin brother
living in the mountains,
ages quicker
than his brother
at sea level,
and by my own
calculations,
I am now
three years senior
to my older brother.
Furthermore,
it's not
the height,
it's the mass.
Planets
orbiting our universe,
cause space
and therefore time,
to bend
around them,
like a bowling ball
dropped
on to a trampoline,
and the further

we are
from a planet,
the faster
time moves
around us.
Time
can't be told
in years,
it reveals itself
in light,
as our sun
sways,
as our planet
spins,
in a universe
ebbing and flowing,
flexing and curving,
like the surface
of the sea.
It becomes
apparent that
it's not
the years,
or the miles,
brother,
it's the mass.

QUANTA

are molecular
dropouts.
Full of potential,
they are lumps
of energy
that materialize only
through interaction
with another,
through "quantum leaps"
they exist
otherwise,
they are left
in a place
between reality
and imagination,
is
and
isn't,
undisturbed,
sitting
cross legged,
eyes glazed,
eating
Cheetos.
They aren't here
until they are,
they can't
become
until
they
fulfill
their purpose.
Quanta
doesn't

exist yet,
like myself,
before my father
found my mother,
before my voice
learned to sing,
before my fingers
touched the strings,
and before
I heard
you speak my name.

MAN ON THE MOON

Man on the moon

I believe
Neal Armstrong
walked
across
the moon
and I also believe
that part
of him
is still there,
never
returned,
because once
he glimpsed
the vastness,
the calm,
the complete
stillness
of space,
it showed him
his eternal nature,
and when
a man

is confronted
with a thing
like that,
he either
runs screaming,
or pulls up a chair.
If there is
a man
on the moon,
I'm sure
he is sitting
comfortably
in his
favorite recliner,
waiting
patiently
for the next woman
on the moon
to pay him
a visit
and give
him a reason
to make
his bed.

Clouds

Our universe
was born
from one
small,
angry cloud
and has been growing
ever since,
is expanding
as we speak.
Astronomers
have measured
the clouds
between the stars,
called Nebulae,
and determined
that our favorite star,
the Sun,
is one tiny speck
in an endless ocean,
our Galaxy,
and every direction
we look,
we see
an endless array
of dots,
each dot
containing
yet another
galaxy,
with planets
and peoples
and stars
of their own
to rotate around,

and worship
in whichever manner
they see fit.
Maybe,
instead of the Sun,
or a God,
they pray to
the fluffy
white cloud,
passing overhead,
shaped
like a dragon,
a cannon,
an arrow
fired by cupid,
from which
our universe
was born.

QUANTUM MECHANICS

I watched
the lake
like a woman
I couldn't
resist,
hoping
she didn't
catch me looking,
hoping
more
that she did,
and
like a beautiful woman
the lake was poised
on the surface,
while beneath
internal dialogues,
to do lists,
horoscopes,
and political deadlines
moved constantly,
countless
separate
particles,
each particle
made of atoms,
each atom
made
of a nucleus,
surrounded
by electrons,
filled with
neutrons
and protons,

made
of quarks,
and the gluons
hold the quarks
inside.
All of these
in constant
motion,
a closer look
at what
you imagined
to be
a calm lake
tells
the true story.
I fell
from the cliff,
the same way
I fell
into her love.
I dove down
into
the swirling lake,
hoping to sink
far enough below
the surface,
to find
a peace
in which
I could rest,
and ignore
the waters
overhead,
boiling
like the sea.

INERTIA

Inertia

The boulder
rolled
from where
it had sat
for centuries,
perched atop
a column
of stone,
gazing across
the canyon below,
the stream
running through,
the sagebrush
strewn about.
The boulder
fell
and splintered
the centuries
across
the canyon walls,

showering
everything beneath
with the grace
and patience
of a man
who sat
for centuries,
silently
thinking
clearly,
and finally
remembering
how to fly.

HUBBLE TELESCOPE

He told us
he would find
a woman
to escort him
to Spain,
to view
the next
solar eclipse.
He didn't
have a date
yet,
but would,
he informed us
during
an Astronomy 101 class I attended
at
Georgia
Southern
University,
the same professor
who
had been called upon

by NASA
for his expertise
in repairing
the Hubble Space Telescope.
When I was a child,
I believed
that magic
swirled
in the air
around me,
the spaces
between
were filled
with magic,
and all I
had to do
was focus
my energy,
reach out,
and grab
a handful
to manipulate
as I
saw fit.
Einstein
agreed,
he told us
space
is no different
than matter,
they are both
made
from the same particles
that make us
so,
I am connected
to the "space"
around me,

each particle
carrying
a tiny piece
of myself.
Scientists
spent years
of their lives
hoping to see
a proton
disintegrate
and transform
into an electron,
as they will,
but found
that when observed,
the proton
never changed.
What they
didn't know,
is that magic
only appears
when you aren't
looking,
like the photos
the Hubble Telescope
incidentally
took
of the corner
of our cosmos,
deeper,
darker,
further
than ever
imagined,
which revealed
more of our universe
then we
had ever dreamt.

. . . and other

THE SOUND BIRDS MAKE

The sound birds make

I've heard
the sound
birds use
to warn
of fire,
but
I need
a word
for fire
that doesn't
burn,
and
I need
a word
for love
that isn't
no.
My pack
was heavy
upon my back
when I saw

the first
of the
wildfire
flames,
with nowhere
to run
I tied
a red bandanna
over my face
and walked,
looking
for a safe place
to sleep.
In my search,
I stumbled
onto the place
a lost balloon
had landed,
after
it drifted
from the
fingertips
of a dismayed
child,
who cried
for what
she hardly knew,
but
it wasn't goodbye,
it was only
so long,
as the balloon
drifted
free
from the
children,
free
from the city,

across
mountain ranges,
over
the wildfires,
to settle
off trail,
in this
quiet canyon,
where I
came upon it,
as I searched
for a new word.
I folded
the broken balloon
and carried it
in my pocket,
as the smoke
collected,
and
filled the valley
below.

My Grumbling Stomach

My Grumbling stomach

tells the story
of a boy
who went to bed
without supper,
of hard bellied men
whom sailed
frozen seas,
straining their eyes
for land,
as their
supplies dwindled,
their dreams
turned
to fattened calves,
and dark-haired women.
It tells the story
of Icarus,
flying too close
to the sun,
but forgetting
to snack
before takeoff,
he,

along with his blood sugar
crashed,
and now
we sing his praises
as my stomach
grunts and groans,
like a man
on a prison chain gang,
struggling
to raise the pick
above his head,
staring
at the ground
in front of him,
praying
for rain,
as a single
bead of salted sweat
rolls
from his forehead,
and drops down
to land silent
upon
the cracked earth.
My stomach
tells the tales
the bards
have forgotten
and minstrels
won't touch,
in five languages
it speaks
of love,
loss,
the mysteries
of the world
and hey,
are you going to finish that sandwich?

SLICK FIFTY

The scent
of chlorine,
hanging
in the air,
carried me
to Highlands Texas
in the summertime,
to tall grass
and chigger bites,
oppressive humidity,
and Mr. Billnoski's
pool house,
where
I would paddleboat
around the lake,
while my grape
snow cone
melted,
and
dripped
down my leg.
It took me
to Ocean City Maryland,
to time share
heated pools
I fearlessly flung
my young body into
on grey mornings,
filled
with peanut butter
and jelly sandwiches,
too much television,
fresh clean sheets,
and splinters

in my bare feet
from chasing
the pelicans
down the boardwalk.
The wind shifted
and I
could smell
Knoxville
in the Fall,
I could see
the leaves'
colors changing,
and I wanted
to return
to life
as a river guide,
standing
on the banks
of the Pigeon
in late September,
when the tourists
would leave
me in silence,
to say goodbye
to the rubber boats
that had carried me
through the summer.
I drained
the life
from them
and put
them away
beneath
a tin roofed shed,
where inside
the boats could dream
of summer storms
that chased away

their customers,
as the river swelled,
overflowing
the banks
with muddied water,
overcoming
their prison shed,
washing
the boats
into the river,
carrying them
down stream,
and releasing them
from lives of servitude.
The boats
would be free
to sink, or swim,
to stay
or leave,
to fail
or succeed,
as we all are.

COVERALLS

I wish
I was bigger,
so I could fit
into
my Grandfather's coveralls,
the ones
my Grandmother
wouldn't let him wear
to the doctor's office,
or when
he drove
his Cadillac
because,
she'd say,
it made him
look like a bum.
He is gone now
and I want
to wear
those light blue coveralls
with the grease stains
on the knees
everywhere I go,
gliding through
my days,
sleeves rolled up,
waist buckle fastened,
humming the tune
my grandfather hummed
on the day
we planted
green beans
in the garden.
I wish

I was sitting
in the back
of my grandpa's Cadillac,
Mills brothers
singing softly
through the speakers,
as he navigated
across
the icy roads,
across
the Mid-west,
through
the dead of night,
until our family
was safe at home,
asleep in our beds,
and I was waking
the next morning
to my grandfather
adjusting
his hearing aid
in the hallway.

PEN VERSUS SWORD

Pen vs. sword

Whomever said,
the pen
is mightier
than
the sword,
has never
been
in a sword fight.
Whomever
said,
the truth
shall
set you
free,
has never
been
to a
court of law.
Whomever
said,
an apple
a day
keeps
the doctor

away,
was not
a licensed
medical
professional,
and
whomever
said,
practice
makes perfect
should
be jabbed
with
a pen.

Rain barrel

You could
pick
a switch
from
the nectarine tree,
or
a tomato
from the garden.
There was
a rusty barrel
that told
how much rain
had fallen,
darling
call me
darling
squeaked softly
from the
portable
A.M.
radio,
a place
wherein
the attic
was forbidden,
and only a string
separated
a child
from heaven
and hell
as
football helmets
cracked
in the backyard,

in a game
of one on one
that
no one
ever won.
The horses
stamped impatiently
across
the barbed wire,
because
they wished
to be included.

One story home

The Pinyon Pine,
standing
alone
in the desert,
pulled me
towards it.
I traced
the rocks
like the water
from the sky,
like veins
from the ground,
it reached for me,
and
it held me,
proud,
as it grew
above
the surrounding brush.
I wondered
if the sage beneath
the pine
thought it arrogant,
or just a bit much,
and reasoned
that it should live
in a
one story home
like the rest
of its' neighbors.
I placed
my hands
on the trunk
of the tree

and thought
about the nights
the tree
stood,
alone
in the desert,
as I
stood,
alone
in the city,
as we both
thought about
the stars,
and how everything
would be okay
if they
could shine
that bright
every night.

NERVOUS WOMEN

To all
the nervous women,
in their yellow
afternoon
sun dresses,
sitting
inside
Indian restaurants,
sipping
sweet,
hot
tea,
waiting
for me
to appear,
and cause
their
dainty
teaspoons
to swirl,
and
clink
against
their cups,
and
here's
to the smile
that could have
bridged
the gap,
and the hand
that would have
taken theirs
for

as long
as they
wished.

THE OCEAN ISN'T BLUE

it is grey.
The sky isn't grey,
it is everywhere.
I am treading
through it.
When the lights
go out,
I'm not sure
who will
turn them back on.
The ink
will smear,
the tide
will pull,
and
the earth
will spin.
Walking,
running,
eating,
sleeping,
nothing
will
make it stop.
It won't
change the color,
or the pattern,
the matter,
the size,
the weight,
or the shape,
it will still be
grey,
until you choose

a different shade.
I spread
all the water
in the ocean
across
my kitchen table
and imagined
there was an escape.
I sailed
across the sea,
landed
on a new continent,
and burned it
to the ground.

FRONT RIGHT BUMPER

My pickup truck
is sexy,
she told me so,
even with the dings
and the dents
on the front right bumper,
and rear left fender.
She glides
through the traffic,
steadily climbs
the hills,
and roars into action
without hesitation.
The dents,
and the dings
on my truck
will never be fixed,
like the scar
above my left eye
will remain,
like the scar
across
my right palm
and
left shoulder,
front right bumper
rear left fender.
I've driven my truck
in a former life,
that's why
I fit so well
into the driver's seat,
and my hand
rests so easily

upon her steering wheel,
as we spin
off the road,
and speed past
police officers.
She understands
when I need to leave
and never
kicks me out
of her bed,
so I don't mention
her dings,
or her dents,
and she pretends
not to notice
my flaws.

A LETTER TO TODD SNIDER

I wrote a letter
to Mr. Todd Snider.
I can't be certain
that he
will ever read it,
the only thing
that's certain
is
I wrote him one.
I figured
if it worked for him,
and Jerry Jeff Walker,
then
it could work
for me.
He'll get a laugh
as he reads
my offer
to roll his smokes
and change
his guitar strings
before
he tosses it
into the waste basket,
or picks up the phone
to give me a call.
I wrote it,
I mailed it,
and I crossed my fingers,
because an email
doesn't carry the weight,
or the severity
of this situation,

or these
double wide blues.
Amen.

STEPS

Steps

I can't
face it today,
any of it.
The sunshine,
my alarm clock,
my dog,
waiting patiently
to be walked
all seem
impossible,
insurmountable,
unattainable.
I don't know
how I'll summon
the strength
to leave
my bed,
pull
open
the blinds,
walk down
three flights
of stairs
and face
this day.

It's
a good thing
I lied
about the
steps.

PILLOWS

We share
the small
orange
pillow.
Sometimes,
it's my
lower back
couch
cushion,
sometimes
it's my dog's
dream cradle.
There's
nothing worse
than
needing a pillow
and not
having one.
A folded jacket
just isn't
the same.

SADDLE

The cheap
hotel soap
stuck to my skin
like
a filthy aftertaste.
The Northern winds
blasted across
the High plains,
feeding
the California fires,
that were devouring
the Californian homes,
tossing
tractor trailers
like rag dolls
off the interstate,
and into
each other.
I held my breath
passing them,
trembling,
as the hand of God
shook my truck,
I waited
for the wind
to pull the machines
down
on top of me.
My doubts swirled
with the dust,
but my truck
knew the way.
It carried me,
a wounded man,

slung
across his saddle,
eight hundred miles,
home
to bed.

YESTERDAY'S LIST

I'm not going
to make
another list
this morning.
I won't look
at yesterday's list,
smugly
hanging
on my refrigerator,
believing
he is victorious.
What yesterday's list
doesn't know,
is
I have made
a new list,
an invisible list,
a secret list,
and
the first thing
on the secret list
is to avoid
all lists.
Check.
Damn.

BIG AGAINST THE SKY

The day seemed
so big
against the sky,
I just wanted
to go back home
but
not home here,
home to Georgia,
but
not home to Georgia,
home in Texas,
home inside
my changing
memories,
and that safe
warm place
in my mind
where
I am free
from harm,
and no one
can touch me,
because
I am surrounded
by a fortress
of memories,
built to protect me,
and carry me
through the days
when the wind
stops blowing,
but
my face
continues to burn.

INCOMPLETE

In complete

I apologize,
but I could not
complete
the assignment.
There was
a great flood
that washed
all the stationary
away
and ruined
every computer
in this
zip code.
I really wanted
to finish
my homework,
but a hurricane
blew in
through
an open window
and scattered
my papers
across the four corners
of the Earth,

where they continue
to float,
and toss,
and turn,
like a guilty man
on a warm night.
Forgive me
for not completing
the assignment,
the stock market
crashed suddenly
and the weight
of the recession
that followed
pulled the ink
from my pages,
and
I couldn't possibly
finish my thought
because
the Mayans
predicted
the collapse
of the planet,
and I had to focus
all my energies
on prayer,
love,
and enough coffee
to keep
us going.
I am truly sorry
for not finishing,
but I couldn't
find my pen,
my garbage disposal
was making
a funny noise,

swarms of locusts
appeared
in my living room,
my pages
were engulfed
in flames
by the Holy Spirit,
and then
my dog ate them.
My sincerest apologies
for not com. . . .

HIDE

We lie in wait,
we do not
make a sound.
We lie still,
our bodies
absorbed
by the night.
We can't
be bothered
by the flickering
of the fireflies.
We do not
need to close
our eyes
to see
the dream shapes.
We hold
our breath,
we do not
make a sound.

SHY

Shy

I'm not shy,
so when
my eyes
look
to the ground,
it's not because
they don't
want
to see you,
it's because
they are afraid
the Love
hiding behind
will come
bursting through,
ejecting them
from their sockets,
which
would be messy,
and
leave me
blind,
to grope
my way
awkwardly
towards

you
through
a blanket
of darkness
for the rest
of my life.
That's all.

JEFF BUCKLEY

said
that it's never over,
my kingdom
for a kiss
upon her shoulder,
but how do you know
when it's never over,
and how many times
should you allow her
to place your belongings
in a pile
in her driveway.
I'm depending
on your ancient wisdom
to see me through
Mr. Buckley,
I've been hardened
by the flames
of passion
and a woman's scorn,
now I lay here
petrified.
I need your help
Jeff Buckley,
send down
your armies
to fight the chaos
in her mind
and the loneliness
in mine;
all my blood
for the sweetness
of her laughter
would leave me

lightheaded,
so I would settle
for a smile
instead.

LAWN CARE

Lawn care

I woke
to the sound
of machines
and prayed
for an earthquake,
flood,
Armageddon,
Black Friday,
Black Elvis,
or anything else
that would make
the weed whackers
stop whacking,
the lawn mowers
stop mowing,
and the table saws
to stop buzzing
through my open window
at seven
thirty
A.M.

on a
Saturday morning.
My neighbors
have forgotten
about entropy,
everything
will eventually
slide
into decay
so
someone should tell
the gentleman
with the weed whacker
that he
is not preparing
to jump across
the Grand Canyon,
and
he isn't Evil Kanevil
so please,
stop revving
the hedge trimmer
like it's
a motorcycle.
I'm not sure
when my neighbors
held their meeting,
deciding to begin
their noises
simultaneously
outside my window
at seven
thirty
A.M.
on a
Saturday morning,
but their journey
to Home and Garden

magazine
began today,
so tonight
I will take
all
of their mowers,
whackers,
hammers,
and hoses,
and throw them
off a cliff,
so if anyone
does happen to find them,
and decide
to trim their grass
early on a Sunday morning,
then at least
there will be a mountain
separating them,
from my bedroom
window.

LAZARUS

What
was I thinking
dating
a Sagittarius
and what
was she thinking
when the sun
began to set,
and she
sat
there
silently
on my hardwood floor,
tired,
unable
to connect
the dots
she had followed.
She sat,
unravelling
like a ball
of yarn,
waiting
for my hands
to give her shape,
but my fingers
were too sore,
so she
laid there
on the floor,
close enough
we finally touched,

and the boredom
was enough
to kill us both.

Parking permit

Parking Permit

The squawking birds
have traded their home
in the trees
for a home
beneath my car port,
but they
don't realize
they must
have a permit
to park there,
otherwise,
I will have
them towed
I say,
as they flock,
swarm,
and
encircle me
with their cries.
Together,
we are

lifted above
the city
and we
drift away,
the birds
surrounding me,
their tiny flapping wings
holding me aloft,
won't let me go,
as I pity
the people below,
going
about their day,
and sincerely hope
they brought
an umbrella.

FRUIT FLIES

The fruit
I buy for decoration,
along with the bread
that never gets eaten,
and will linger
alongside the bills,
and the rent,
as the fruit flies
collect
in
my mind,
until
a thick buzzing cloud
separates me
from my thoughts,
until the muddy paw print
on the newly swept floor
is a tidal wave
building in the distance,
as the ocean
pulls at its' shores,
the breath
is taken from me,
my insides
begin to turn
out,
I am tossed
into the ocean,
become the ocean,
always was
the ocean,
waiting
for the clouds
to call me home.

I rage
and foam,
while an empty bottle
bobs quietly
on my surface.

Nightmare Berets

Nightmare Berets

It's the thought
that sneaks up
and lays
across my chest
in the middle
of the night,
when I awake
from tossing
and turning,
because
it was
too hot,
too bright,
too loud,
too scary.
I throw
the thin
damp sheet
from my
sweaty body,
and the fears
are there
waiting,
climbing

the covers,
daggers
between their teeth,
French berets
tilted provocatively.
They whisper dread
into my ear
and ask
why
am I alone,
what
have I done
with my life,
and when
will I do something
with myself?
They hold me down,
strangling my dreams
until
the rising sun
chases them away,
but as I
go about my day,
the fear,
now
a distant memory
in the
morning light,
waits for another
sweaty,
restless,
toss
and
turn,
clench
your teeth,
straighten
your legs,

sleepless
night
to pay me
another visit.

ATHENS

Many
are the miles
that separate us now
and long
were the legs
that carried her away.
Longer still,
are these nights
and the dreams
that return
again
and again
to that bar room,
and those strange characters.
I couldn't
figure out
why the homeless man
I passed
on the sidewalk
said, "Don't worry,
it will all
be over soon,"
and I wonder
how
he knew.

ROCKET SHIP

"Well, I never"
she said,
but if she
ever
did,
she would
do it
in a rocket ship,
hot pink,
caked in glitter,
shining
like the sun,
blasting off,
headed
straight or
crooked
depending on
wind currents
and
the latest
gossip.
If the girls
who *never*
had tried
ever
then never
wouldn't be
so far away
and
wouldn't require
a rocket ship
to reach.

Life Lessons

Coyote bones

I spent the night
lying
in the dirt,
curled
around a fire,
counting
my bones,
and wondering
when
I had grown
so thin.
The breeze
lifted my scent
and carried it
across the desert,
bouncing
off the rocks,
the sage,
and Pinyon pine,
as
the animals
licked
their curling lips,
and drooled
over the
delicious smell
hanging in the air.
I laid
where the coyotes had,
beneath
the Pinyon pine,
where
they had scattered
the rabbits' bones.

I gathered
their remains,
shook them
in my hands,
and spread them
on the ground
to see
my future.
The bones
showed me
days with sunshine
and days of rain,
there would be times
when I was lonely,
and times of joy.
I would find love
and I would lose love.
The bones
told me to breathe
softly
and step softer,
to listen to the wind,
follow the rivers,
and most importantly,
they warned me
to never
trust
a coyote.

SINK

My sink
is two years
older than me.
I know this
because the date
4-4-77
is written
in black marker
beneath its' basin,
and I know this
because
my cupboard doors
no longer
cover
their cupboard,
they fell off
during a night
of unprovoked sink violence,
and I never
saw the need
to replace them.
Now,
I always know
when I am running low
on toilet paper,
ibuprofen,
or
triple
antibiotic
ointment.
My sink
is two years
older than me
and like myself,

is still looking good,
and doesn't drip
too often.
It teaches me
life lessons
with its' silent wisdom
and I
pretend
to listen patiently
to what it would say,
imagining
the advice
it would give,
in a soft,
scratchy voice,
choked by years
of hair,
smoke,
love,
and rust.

GUARDIAN ANGEL

Guardian Angel

My guardian angel
must be
just as confused
and aggravated
as my
high school
soccer coach was,
baffled
by my
poor decisions
and
constant efforts
to blow it
just when
the game
is within reach.
How frustrating.

NINE SCOOPS

This coffee
tastes weak.
Nine scoops
went into
the pot,
boiling water
was poured over,
but
it brewed
reluctantly,
and continues
to procrastinate
as it sits
here
muddied,
confused,
lying
in my
favorite mug.
Next time,
ten
scoops.

Alcoholics Anonymous

When life
gives you lemons,
make
a gin and tonic.
When gin
and tonics
give you cirrhosis,
make it
to an A.A. meeting.
When A.A. meetings
give you gin
and tonics,
find a new meeting.

Radiant

When I smell smoke
in the air
I think of her,
that tangled
beautiful mess,
her crimson sheets
thrown across
an unmade bed,
and when
the summer rain
falls over
the wildfires,
like her
yellow hair
fell over
me,
I wonder
how I ever
did with,
or without her.
Curled together,
her glitter
would flake
onto my skin,
when the light
caught me right,
I was an angel
translucent and shimmering.
Lying together
we were safe,
but beneath
her bed
something
waited for us,

something
we couldn't name,
and didn't speak of,
but
we had practiced
our lines
so many times,
goodbye
was effortless,
seamless,
and I floated
from her house
into the night
an angel,
glowing radiant,
sparkling
with glitter,
love,
and hate,
searching
for someone
to pray over.

SHIPWRECKS

I've left the comfort
of the covered porch,
the gentle gazebo,
the placid place
where the fountain
drowns out
the sound
of airplanes
cars,
and the comfort
of the sun
pulls me in closer,
tells me everything
will be alright,
just sit still,
and let me
warm you.
The clouds
aren't convinced,
they lie
in wait
for a new day
when they are king.
If the wind
would ever stop
blowing,
I could catch
my breath,
reposition
my heart,
and ready my feet
for travel.
I've been still
for too long

and the restless
twitch
of my hands
tell me
it is time
to start
moving
them again.
I'm not sure
in which direction
to turn.
The sun says East,
the stars West.
The wind swirls
in circles
and my thoughts
tumble
down a gully,
headlong
over a cliff,
swept downstream,
washed ashore,
I have time
to shoot
a quick bearing,
to triangulate
by heaven and hell,
and pinpoint
my location.
It's not
where you are
that's important,
it is
where you are going
that matters,
or
is that backwards?
It's best

not to trust
a desperate man,
he will reach out
for help,
only to pull
you down with him.
The next time
you see
a drowning man,
throw him
a life preserver
and keep rowing.
Monuments are built
from shipwrecks,
but these
aren't statues
I want to visit.

AIR CONDITIONING

My body
pulled me
outside
into the
humid,
Oklahoma,
eight-year-old,
summertime
heat,
where the sun
could flash fry me,
sear my skin
until
I could no longer
endure it,
and was forced
by the scorching sun
back inside,
into the stale,
cool
air-conditioned air,
where Grandma
washed the dishes,
and Grandpa
napped
through
the afternoon heat.
I would try on
my uncle's
cowboy boots,
or my
Grandfather's hats,
until I grew tired
of waiting to grow

into them,
and the false
cool air
collected
upon my smooth skin,
tingling
with the lies
of modern day,
and I
would run
back outside,
to let
the sunlight
melt them
away.

Bus

If a seventy-two
passenger bus
came careening
around the corner
and flattened me
like a pancake,
I wonder
if I would
still
be thinking of her
at the moment
of impact,
or
if I would notice
the model,
or the make
of the tires
crushing my bones,
and ending
my existence.
Would I still
be picturing
her smile,
or
would I think about
how dirty
the bus's windshield
was,
and ponder
how the driver
could operate
the vehicle
from behind
such filthy glass,

with headlights
blinding,
shining through,
highlighting
the final remains
of thousands
of insects,
who were thinking
of sunlight,
and picnics
before
they were annihilated.
Would I think
of the sound
of her voice,
or
would I be distracted
by the screech
of bus's brake pads,
grinding
metal on metal,
sparks singing
their own chorus,
because
every sound
carries a note,
and each note
is heard by someone
somewhere,
perhaps
by the distracted man,
standing
in the middle
of the road,
daydreaming
of lost love.

My coffee pot is nervous

My coffee pot is
Nervous

When I drink
a pot
of coffee
in one morning,
it means
I am anxious,
but
I'm not sure
if I'm anxious
because
I drank
an entire pot
of coffee,
or
if I was anxious
so
I drank
an entire pot
of coffee.
When I wake
in the morning
and find
my toes
curling
into balls,

it means
I'm anxious,
but
I'm not sure
if I
wake anxious
so I curl
my toes
into little balls,
or
if I
become anxious
because
I am curling
my toes
into
anxious
little balls.
Phew,
it is exhausting
being so on top
of my emotions.

A Poem for a frightened adult

Have I told you
of
the dark grey cloud
that hangs
over my head?
Have I spoken
of
the monster
that lives
under my bed?
He reaches up
to grab my leg
when I throw
my covers off,
he cannot sleep
unless I'm awake
thinking useless thoughts.
Did I mention
all the times
I've tried
and never
seemed to make it,
or all the nights
I've given up
like it didn't matter?
Have I told you
of
the dream
I had,
to help me
guide my way?

I hope
I didn't
name the path,
it's mine
alone to take.

CATERPILLAR

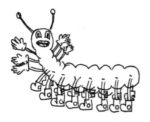

Caterpillar

After two weeks,
the caterpillar
will emerge
from his
silken cocoon,
a triumphant butterfly,
free
from the burden
of dragging
his feet
upon the ground,
he will travel
on the wind,
carrying beauty
and hope
to all
who see him.
Not this caterpillar.
I found
him
when
I was a boy,
age ten.
It looked like
an ideal specimen,
a perfect pet,

a side kick,
he looked like
he
could be my friend.
I emptied out
a plastic bucket and
placed him inside
where he would be safe.
I left him
torn blades of grass
to eat.
He didn't stay long.
His body
dried
and curled
into a ball,
the final remains
of a
silent companion,
leaving me
to face the world
alone,
or until
I decided
to turn over
another stone.

Oklahoma weather

This
is Oklahoma weather,
this is how
the air tasted
on the day
I was released
from county jail.
A cosmic swirl
of fear,
hope,
and hate
gathered above
as I journeyed the
muddied dirt roads
that suck,
slide,
and pull
you out into
black
nothingness.
It's been happening
from the moment
I was considered
and will continue
to happen
past the time
I am forgotten.
Hearts are beating
across the world,
eyes are crying,
breath is taken,
given,
taken,
given,

and all
the answers
lie on the floor,
scrawled
in the dirt,
written
in scars
across
the backs
of those
who have escaped.

CAPTAIN COININGTON

Captain Coinington

sits on the dresser
and watches me
with his good eye.
We've travelled
across the Atlantic,
across the United States,
together
for thirty years.
Some would say
he is a coffee mug,
but I would say
he is the story
of my life.
Dressed
as a pirate,
an eye patch covers
one eye,
his grizzled beard
and scarred face
tell
of his adventures,
and he,
along with a bit of advice
about not eating
late at night,

are two
of the few
things
my father
left me.
Coinington
sits
on my dresser,
laden with the coins
of differing currencies,
live and dead,
old guitar pics,
and occasional
hair ties,
which are
his treasures
to guard
with his life,
or until
I need
to do laundry.
Captain Coinington's
gaze is steady
and reassuring,
he sits
next to me
studying the sea,
navigating
by the stars,
searching for
his ship,
his white whale,
his home,
his escape
from a life
spent
on
a shelf.

Enjoy the Night

Enjoy the night

I don't
name it
so it doesn't
have a face,
eyes to see me,
or lips
to call me.
I reach for it
so I can
swing it
in circles
above my head,
cooling it
before
bringing it
down
to caress
my cheek.
The days
are different
and strange

as
I watch my hands
no longer
clutching
for a handhold
of a sinking ship.
It waits
for me
in the grey clouds
and
restless nights,
a stranger
in a dark alley,
holding a bottle,
and a handful
of promises,
telling me to relax
and just
enjoy the night.

BLUE TIN MUG

I wrapped
my fingers
around
my blue tin mug
and the
steaming coffee
inside
moved through
the thin metal,
seeped into my skin,
was absorbed by
my bones,
and flowed
through my blood.
I was no longer
the thin man
searching,
I had found
my reward
and I beheld
all the treasures
the earth
had masked
in late
afternoon fog.
Lovely,
like a woman
just out of reach,
when you finally
have her,
you're not so sure
what all the fuss
was about.
I tucked

her memory away
and climbed back
into my tent,
to watch
the clouds
bring the rain closer.

ANOTHER POEM ABOUT FLYING

It wasn't a dream,
but a memory.
I jumped
and flew,
barley clearing
the trees
and houses below.
The night air
was cold,
I was afraid,
but I kept flying,
because
I knew I could.
I was born
able to fly,
it has taken me
a lifetime
to remember how.
Our bodies
are lighter than air,
our souls
darker than soot,
our hearts
make the decision
to stay or leave.
Run as fast as you can
and jump,
if you find yourself airborne
push higher,
feel the nighttime
on your face,
lifting
your ageless body,
and keep going.

Don't worry about
packing a lunch,
you can
live off
the light
of the moon,
it will
feed your soul
until
you need to land,
where
I will
be waiting
with a blanket,
pillow,
and your ride
home.

LEPRECHAUNS

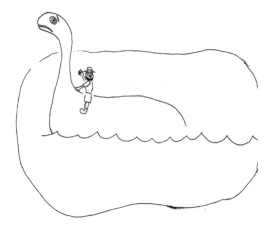

Leprechauns

I knew
I wasn't supposed to,
but the Sun
was shining
and her smile
looked
so real,
I believed
in everything,
all
of it:
Love,
Leprechauns,
Werewolves,
the Loch Ness monster,
the Tooth Fairy,
Democracy,
and Social Security.
I knew
good
and well

what
I was doing
wasn't
good
or well,
but the sky
was
a pure blue,
not
a single
cloud,
fluffy
or grey
to distract,
so I kept
walking
with her,
knowing
each step
would
take me
further
from where
I needed to be.
Let's hold hands.

A PERFECT NIGHT

Tonight
is a perfect night
to visit a cemetery,
the air is still
without a breeze
to disturb
the ghosts.
Tonight
is a perfect night
to rearrange Stonehenge,
create constitutions,
and give nations
new meanings.
Tonight
is a perfect night
for a rain
that never stops,
taking a vow
of silence,
solving the riddles
of the Sphinx,
digging
a hole to China,
wandering
through corn fields,
strapping on roller skates,
exorcizing demons,
and
clipping angel's wings.
Tonight
is for
borrowing ideas,
signing
high school yearbooks,

catching snakes and feeding
them to rats,
lying with thieves,
and watching
the pendulum swing.
Tonight is a perfect night.

ELIOT SMITH

I was thinking
of sunsets
and
beauty,
longing,
love everlasting,
and how hard
a man
would have to drive
a knife
through
his chest
to pierce
his own heart,
to rip through
the tattooed skin,
the blood,
the bones,
the hopes,
and the songs,
and how easily
one woman
can cut a man
down,
with a
short
black dress,
on any
given night.

Technology

It stopped
working,
so I shook it.
I bargained,
I begged it,
I rattled
its' bones
until
the tiny pulse
stopped beating
inside,
still,
I wanted
it to work
like it did
when
first purchased.
I would
touch the button
and the device
would
spring
into action,
bringing me
a glowing happiness
the moment
I asked.
We worked
together
in unison,
the beautiful
machine
and myself,
and life

was
as it should
have been,
held
in perfect
balance
by
the angels
and
innovators
of technology,
but
this piece of junk
stopped working,
so I threw it
against
the
wall.
Let's go shopping.

Chain saws and fickle women

It's my knees sir,
it's always been my knees
buckling, swaying,
threatening
to bend
backwards,
on themselves
if not cared for properly,
and as my knees
snap and crackle,
I wait for the pop
each time I bend.
I wonder
about the oak tree
in my yard
and its' branches,
as they
break
and fall to the ground,
I think
about my own limbs,
and how long
it will take
for them to wither,
and snap
from my trunk.
Much like the rings
around the trunk,
knowledge
lies inside of us,
waiting to return
the wisdom
to the soil
for the next,

so they
can learn,
teach,
and grow tall,
as they reach
towards the light,
taking for granted
they steer clear
of chain saws
and fickle women.

BLACK OUT

When the lights
went out
I spoke
a prayer,
asking
for a forgiveness
I had never shown,
mercy
beyond
my comprehension,
a grace
that would silence
the wicked,
a second chance,
a third chance,
a glimmer of hope,
a spark,
a distant promise
that the light
would one day
shine again,
as I waited
for the sky to part,
and the trumpets
to sound.
When the lights
came back on
I prayed
for a black out.

THE ALAMO

I'm glad
that I
was born
in the great state
of Texas.
It excuses
my arrogance
and explains
why I like
cowboy boots
so much.
It tells me
why
I'm
so fierce,
independent,
and
fiercely independent.
These ocean waves
rolling over me
feel so strange,
I need
to walk through
a desert
and let
the sagebrush
scrub me
clean.
I used to dream
of Davey Crockett
and the heroes
of the Alamo,
standing
with them,

side by side
in battle,
muskets blazing.
In other visions,
I'm a deserter,
afraid,
hiding
beneath piles
of slain soldiers,
feigning death
as the cannons
roared,
and the waves
of Santa Anna's army
washed over me,
with only
the mud walls,
of a crumbling fortress
standing
between us,
deep
in the heart
of Texas.

PHASE ONE PROPHECIES

Fallen
out of love,
falling
from a life
I once knew,
further
than I ever expected,
baffled
by
the distance,
by
how far
behind I am,
I am,
I am,
a powerful being,
who transcends
time,
eludes gravity,
viruses,
and obesity,
whom
laughs
at the
buildings
falling around him,
whose form
can't be defined
by geometric shapes,
whom exists
on a plane
unseen,
impossibly imagined.
I cause

the rain
to fall,
the mountains
to crumble,
the sea
to swallow them,
and return home
to watch
Money Heist
on Netflix.
I could speak
of the mysteries
of love,
the formation
of the stars,
how the sea
learned to swim,
how the fish
grew their legs,
and how
the tiger
got its' tail,
but
I'm currently
focusing
all my energy
on job applications,
and the least
expensive
meal delivery service.
I am
a powerful being.
I can
cause others
to run
from me,
flee screaming,
by removing

a cloth mask
from my
wheezing face.
I can empty
grocery stores
with a single
sneeze,
I can
repel strangers
by offering
my hand,
and
I can
See-saw
in solitude,
provided
the barbed wire
surrounding
the playground
doesn't scratch me
as I scale it.
I can see
the future clearly,
stock values
will rise
and fall.
Babies
will be born,
the elderly
will pass on.
Great crimes
will go unlooked
and great minds
will be ignored
by powerful,
ignorant
white men.
Temperatures

will rise
and banners
will fall,
as this planet
is returned
to mother ocean.
I am
a powerful being,
now please,
give me
six feet
of space.

Visits from strangers

If not now,
then when?
If not
through
these people,
during
this sickness,
through
this panic,
if fathers
don't forget
their grocery store
greed
takes flour
from
their neighbor,
steals bread
from his
daughter,
if they can
remember,
their hopeless
doesn't
mean me.
Now is the time
to realize
despair,
comes from
not having
a task
to complete,
a list
to check,
a name tag

to hold
form
into place,
a title,
to gather
an identity around,
lost,
abandoned
by supervisors,
afraid,
unsure
of what
lies inside,
now
is the time
to find
a creek,
kick off
your shoes,
peel back
your socks,
and walk
barefoot
upstream,
into
the woods,
without
a flashlight
to lead,
a journey
without GPS,
to talk
to the man
with
the beard,
holding
the sign
on the corner,

to consider
religion
was created
by men.
Now
is the time
to find
a hobby,
to explore
your neighborhood,
to leave
the light on,
and hope
a stranger
comes
for a visit.
Now is the time
you've been
waiting for.

SCAPEGOAT

I'll blame it
on the
withdrawal
symptoms,
not
the rising
cloud of dust
stirred
by the
crumbling
of my faith.
I'll blame it
on the man
sitting
in Office,
our depleted economy,
computers
that enslave,
blinking screens
that captivate,
and winding circles
that grind
your minutes
away,
as you stare
waiting,
waiting.
I'll blame
the bad days
on Facebook,
as well
as the good ones.
I'll blame
the rain

on the
weatherman,
the
weather app,
the
weather
conditions
on Pluto,
or any other
planet
that never existed.
Blame me
for the height
of my lawn,
covering
my house,
my shame,
casting
a shadow
over my neighbors,
because
I am leaving it
as it lies,
allowing
the weeds
to grow,
and
the sprinklers
to fail.
I am walking
away
from the madness,
where
my footsteps
leave silent
trails
in the dirt,
traces

in the sand,
I'll hike
into the forest
lay upon the moss,
and rest,
relieved
there is no one
left
to blame
but
myself.

STREETLIGHTS

Streetlights

A moment
of silence
for Guy Clark,
Scott Weiland,
Chris Cornell,
Prince,
and Tom Petty.
A moment
of silence
for Billy Kaiser,
Rico Hood,
and all
the broken hearted
sinners
out
in the pitch black,
no radio,
middle America
corn field nights.
For the writers
sitting slouched
in bars,
whose words

have never been read
by the one
they wrote
them for,
dreaming
of sobriety,
or success,
and with
each sip
distance themselves
further from it.
A moment of silence
for the artist,
staring
at his blank easel
in disbelief,
unsure
of how
to pay the rent,
and one final
moment of silence
for the children
we once were,
whom played outside
as long
as we could,
until
the streetlights
glowed,
telling us
it was time
to come home.

CPSIA information can be obtained
at www.ICGtesting.com
Printed in the USA
FSHW022001171121
86166FS